Md. Lutfar Rahman

Basic PLC and Elevator Control System

Md. Lutfar Rahman

Basic PLC and Elevator Control System

A Very Easy Way to Understand PLC and Elevator Control System

LAP LAMBERT Academic Publishing

Imprint

Any brand names and product names mentioned in this book are subject to trademark, brand or patent protection and are trademarks or registered trademarks of their respective holders. The use of brand names, product names, common names, trade names, product descriptions etc. even without a particular marking in this work is in no way to be construed to mean that such names may be regarded as unrestricted in respect of trademark and brand protection legislation and could thus be used by anyone.

Cover image: www.ingimage.com

Publisher:
LAP LAMBERT Academic Publishing
is a trademark of
International Book Market Service Ltd., member of OmniScriptum Publishing Group
17 Meldrum Street, Beau Bassin 71504, Mauritius

ISBN: 978-3-659-62215-1

Zugl. / Approved by: Shenyang, Shenyang University of Chemical Technology, Diss., 2014

Copyright © Md. Lutfar Rahman
Copyright © 2014 International Book Market Service Ltd., member of OmniScriptum Publishing Group

Acknowledgement

Foremost, I would like to convey my sincere gratitude to my advisor Mr. Xu Chengtao for his patience, inspiration, motivation and enthusiasm. His immense knowledge and guidance helped me to complete my project successfully. He is not only the best advisor but also a committed mentor for my study.

Besides my advisor, I would like to thank my friends and teachers for their encouragements and continuous effort.

Furthermore, I would also like to thank my beloved parents Mr. S.M. Mazibur Rahman and Mrs. Lutfa Akter for giving birth to me at the first place and supporting my studies. From my bottom of my heart, I pray for them to be healthy and safe rest of their life.

Finally, special thank to "LAMBERT Academic Publication" to grant my project and also thank them for publishing it carefully.

Md. Lutfar Rahman

Basic PLC and Elevator Control System

Table of Contents

Chapter 1 Introduction 4
1.1 About PLC 4
1.1.1 PLC History 4
1.1.2 Advantages of PLC 5
1.1.3 Basic Operation of PLC 5
1.1.4 PLC programming Language (Ladder Logic) 7
1.1.5 Selection of the PLC Model 7
1.1.6 Why do I use Plc Device 8
1.2 About Elevator 8
1.2.1 History of the Elevator 8
1.2.2 Principle Structure of the Elevator 9
1.2.3 System of the Elevator 11
1.2.4 The Main Parameters and Size of an Elevator 12
1.2.4.1 The main parameters of the elevator: 12
1.2.4.2 The size for the standard elevator: 13
Chapter 2 Request of the Elevator Control System 14
2.1 Control Requirements 14
2.2 Process Solutions 15
Chapter 3 Hardware System Configuration 16
3.1 Elevator System Design 16
3.2 PLC Selection 16
3.2.1 Overview of Siemens S7-200 PLC 16
3.2.2 PLC and its Application Characteristics of the Elevator Control 17
3.2.2.1 PLC Features: 17
3.3 The Idea of PLC Control System Design 18
3.4 The Main Research Content 20
3.4.1 PLC Input / Output Point Estimation 20
3.4.2 The Estimation of PLC Memory Capacity 20

3.5 PLC Input / Output Assignment	21
3.6 PLC Control Circuit Design	22
3.7 Main Circuit Diagram	23
Chapter 4 Software Design	24
4.1 Program Design	24
4.1.1 Program Flowchart	24
4.1.2 Flowchart of the Elevator Up and Down	26
4.1.3 Ladder Logic Diagram:	27
4.2 Program Debugging	36
4.2.1 The Hardware Debugging	36
4.3 Software Debugging	36
4.4 Edit and Compile	36
4.5 Download	36
4.6 Monitoring Program and Running the Debug	37
Chapter 5 Conclusion	38
5.1 Check	38
5.2 Summary	38
Appendix A – About S7-200	39
Understanding How the S7-200 Executes Control Logic	39
Appendix B – About Ladder Logic Diagram	43
What is Ladder Logic	43
What advantage is there to programming in ladder logic?	43
References	46

Chapter 1 Introduction

1.1 About PLC

A Programmable logic controller(PLC) is a digital computer used for automation of electromechanical process, such as control on factory assembly lines, control of amusement rides, or control lighting fixtures, motors control, elevator control and so on. A PLC monitors inputs, makes decisions based on its program, and controls outputs to automate a process or machine. This course is meant to supply you with basic information on the functions and configurations of PLCs. Plc have developed to replace relays.

1.1.1 PLC History

Before the PLC, control, sequencing, and safety interlock logic for manufacturing automobiles was mainly composed of relays, cam timers, drum sequencers, and dedicated closed-loop controllers. Since these could number in the hundreds or even thousands, the process for updating such facilities for the yearly model change-over was very time consuming and expensive, as electricians needed to individually rewire relays to change the logic.

"Digital computers, being general-purpose programmable devices, were soon applied to control of industrial processes. Early computers required specialist programmers, and stringent operating environmental control for temperature, cleanliness, and power quality. Using a general-purpose computer for process control required protecting the computer from the plant floor conditions". An industrial control computer would have several attributes: "it would tolerate the shop-floor environment, it would support discrete (bit-form) input and output in an easily extensible manner, it would not require years of training to use, and it would permit its operation to be monitored". The response time of any computer system must be fast enough to be useful for control; the required speed varying according to the nature of the process.

In 1968 GM Hydra-Matic (the automatic transmission division of General Motors) issued a request for proposals for an electronic replacement for hard-wired relay systems based on a white paper written by engineer Edward R. Clark. The winning proposal came from Bedford Associates of Bedford, Massachusetts.

The first PLC, designated the 084 because it was Bedford Associates' eighty-fourth project, was the result. Bedford Associates started a new company dedicated to developing, manufacturing, selling, and servicing this new product: Modicon, which stood for Modular Digital Controller. One of the people who worked on that project was Dick Morley, who is considered to be the "father" of the PLC.

The Modicon brand was sold in 1977 to Gould Electronics, and later acquired by German Company AEG and then by French Schneider Electric, the current owner, one of the very first 084 models built is now on display at Modicon's headquarters in North Andover, Massachusetts. It was presented to Modicon by GM, when the unit was retired after nearly twenty years of uninterrupted service. Modicon used the 84 moniker at the end of its product range until the 984 made its appearance.

1.1.2 Advantages of PLC

The automotive industry is still one of the largest users of PLCs. PLCs have been gaining popularity on the factory floor and will probably remain predominant for some time to come. Most of this is because of the advantages they offer.

    i.   Cost effective for controlling complex systems.
    ii.   Flexible and can be reapplied to control other systems quickly and easily.
    iii. Computational abilities allow more sophisticated control.
    iv. Trouble shooting aids make programming easier and reduce downtime.
    v.   Reliable components make these likely to operate for years before failure

1.1.3 Basic Operation of PLC

There are four basic steps in the operation of all PLCs which continually take place in a repeating loop: Read Input, Execute Program, Diagnostics and Update Output.

Figure 1.1: Basic Operation of PLC

Check Input Status: First the PLC takes a look at each input to determine if it is on or off. In other words, is the sensor connected to the first input on? Then the second input? Then the third and so on.... It records this data into its memory to be used during the next step.

Execute Program: Next the PLC executes your program one instruction at a time. Maybe the program says that if the first input was on, then it should turn on the first output. Since it already knows which inputs are on/off from the previous step it will be able to decide whether the first output should be turned on based on the state of the first input. It will store the execution results for use later during the next step.

Update Output Status: Finally the PLC updates the status of the outputs. It updates the outputs based on which inputs were on during the first step and the results of executing your program during the second step.

"The basic elements of a PLC include input modules or points, a central processing unit (CPU), output modules or points, and a programming device. The type of input modules or points used by a PLC depends upon the types of input devices used. Some input modules or points respond to digital inputs, also called discrete inputs, which are either on or off. Other modules or inputs respond to analog signals. These analog signals represent machine or process conditions as a range of voltage or current values".

The primary function of a PLC's input circuitry is to convert the signals provided by these various switches and sensors into logic signals that can be used by the CPU. The CPU evaluates the status of inputs, outputs, and other variables as it executes a stored program. The CPU then sends signals to update the status of outputs. Output modules convert control signals from the CPU into digital or analog values that can be used to control various output devices. The programming device is used to enter or change the PLC's program or to monitor or change stored values. Once entered, the program and associated variables are stored in the CPU. In addition to these basic elements, a PLC system may also incorporate an operator interface device to simplify monitoring of the machine or process.

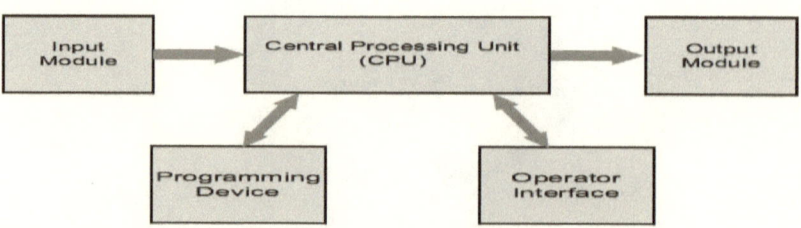

Figure 1.2: Details Process of PLC's Operation

1.1.4 PLC programming Language (Ladder Logic)

In my design I use Ladder logic programming method. A program consists of instructions that accomplish specific tasks. The degree of complexity of a PLC program depends upon the complexity of the application the number and type of input and output devices, and the types of instructions used. Ladder logic (LAD) is one programming language used with PLCs. Ladder logic incorporates programming functions that are graphically displayed to resemble symbols used in hard-wired control diagrams. The left vertical line of a ladder logic diagram represents the power or energized conductor .The output coil instruction represents the neutral or return path of the circuit. The right vertical line, which represents the return path on a hard-wired control line diagram, is omitted. Ladder logic diagrams are read from left-to-right and top-to-bottom. Rungs are sometimes referred to as networks. A network may have several control elements, but only one output coil.

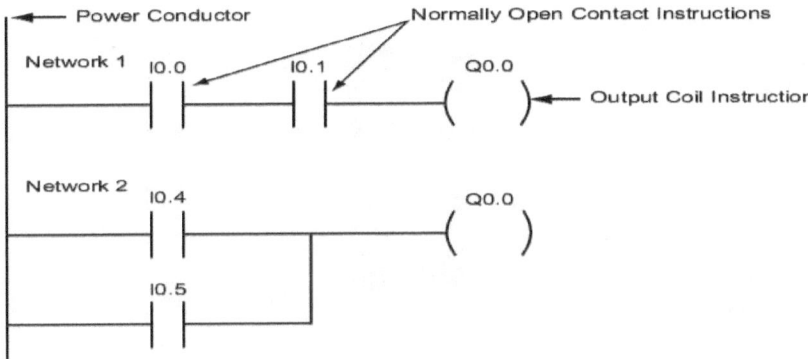

Figure 1.3: Ladder Logic Diagram Section

1.1.5 Selection of the PLC Model

The actual task of implementing automation is therefore paramount, often involves many methodologies in its system design phase. Once the specifications are established, the job of selecting a suitable controller will then become most important as this would determine how at ease the automation program might continue. There is a massive range of PLC systems available today, with new additions or replacements continually being produced with enhanced features of one types or another.

1.1.6 Why do I use Plc Device

PLC is a small computer used for a automation real world process such as control of machinery on factory assembly lines where older automated systems would use hundreds or thousands of relays. A single PLC can be programmed as a replacement and another important thing is it is easy and cheap.

1.2 About Elevator

The elevator (or lift) is a type of vertical transport equipment that efficiently moves people or goods between floors (levels, decks) of a building, vessel or other structure. Elevators are generally powered by electric motors that either drive traction cables or counterweight systems like a hoist, or pump hydraulic fluid to raise a cylindrical piston like a jack.

Figure 1.4: Elevator

1.2.1 History of the Elevator

The first reference to an elevator is in the works of the Roman architect Vitruvius, who reported that Archimedes (c. 287 BC – c. 212 BC) built his first elevator probably in 236 BC. In some literary sources of later historical periods, elevators were mentioned as cabs on a hemp rope and powered by hand or by animals. It is supposed that elevators of this type were installed in the Sinai monastery of Egypt.

In 1000, the Book of Secrets by al-Muradi in Islamic Spain described the use of an elevator-like lifting device, in order to raise a large battering ram to destroy a fortress. In the 17th century the prototypes of elevators were located in the palace buildings of England and France.

Ancient and medieval elevators used drive systems based on hoists or winders. The invention of a system based on the screw drive was perhaps the most important step in elevator technology since ancient times, leading to the creation of modern passenger elevators. The first screw drive elevator was built by Ivan Kulibin and installed in Winter Palace in 1793. Several years later another of Kulibin's elevators was installed in Arkhangelskoye near Moscow.

1.2.2 Principle Structure of the Elevator

The elevator is mechanical, electrical integration products. The mechanical part is like a human body, the human equivalent of the electrical part of the nervous system, the control section equivalent to the human brain. By scheduling the various parts of the control section, in close coordination with the elevator reliable operation. Despite the wide variety of the elevator, the elevator is currently used by the vast majority of electric drive, traction rope structure, as shown in Figure 2-1 is a sectional view of the basic structure of the elevator intuitive map.

The Elevator consists of four parts:

    i. The structure of the machine room,
    ii. Shaft,
    iii. Carrying passengers or cargo space, for the car or passengers or goods
    iv. Landing in or out of the capsule location

The car, cables, elevator machine, control equipment, counterweights, hoistway, rails, penthouse, and pit are the principle parts of an elevator installation.

The Car

The car is a cage of some fire-resistant material supported on a structural frame, to the top member of which the lifting cables are fastened. By means of guide shoes on the side members, car is guided in its vertical travel in the shaft. The car is provided with safety doors, operating-control equipment, floor-level indicators, illumination, emergency exits, and ventilation.

Cable

Four to eight cables, depending on the car speed and capacity, are placed in parallel; in general, each rope is capable of supporting the entire load. The minimum factor of safety varies from 7.6 to 12.0 for passenger lifts and from 6.6 to 11.0 for freight lift.

The Counter Weight

The counter weight is made up of cut steel plates stacked in a frame attached to the opposite ends of the cables to which the car is fastened. Its weight equals that of the empty car plus 40% of the rated live load. It serves several purposes: to provide adequate traction at the sheave for car lifting, to reduce the size of the traction machine, and to reduce power demand and energy cost. Higher initial cost due to strengthen the overhead machine room floor, which must carry the additional structural load of the counter weight.

The Shaft or Hoist

The shaft or hoist-way is the vertical passageway for the car and counterweights. On the side walls are the car guide rails and certain mechanical and electrical auxiliaries of the control apparatus.

The Basic Structure of an Elevator is shown next:

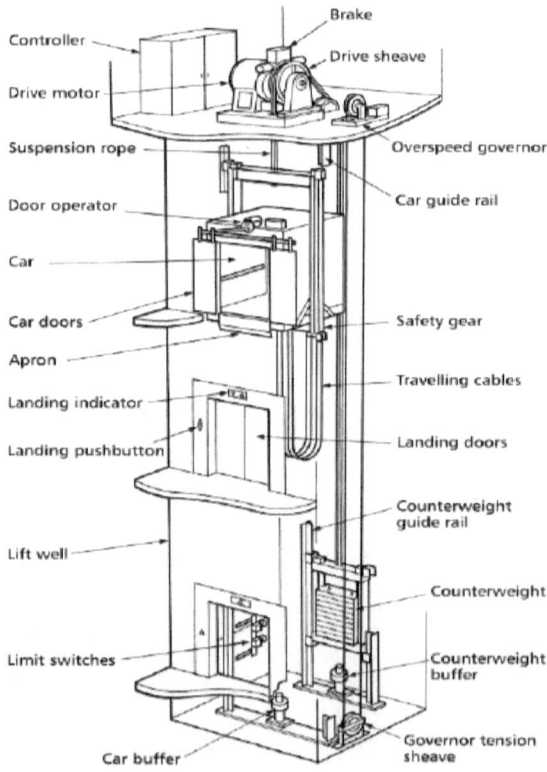

Figure 1.5: Basic structure of an elevator

1.2.3 System of the Elevator

First, drag indexing system

Elevator drag system's main function is the output power and transmission power, drive the elevator running mainly by pulling machine, pulling wire rope, guide wheel and the rope round. Drag power engine for the operation of the elevator is made by electric motors, drag guide wheel, even the axis, reducer, and the electromagnetic brake. The role of the guide wheel is separate capsules and spacing of weight. The process can also be increased by gravity.

Second, the guidance system

Guiding system includes the guide rail, boots and the guide frame. Its main effect is to restrict the capsules and to the activities of the degrees of freedom, the capsules and to only do lifting movement along the guide rail.

Third, the door system

Door system includes the car door, opening the door; open the door, moving and institutions. Capsules includes of capsules entrance door, the door leaf and door guide rail frame and the layer door in standing at the entrance. Open the door machine is located in the capsules, is the power source of the capsules and layer of the door.

Fourth, the capsules

Capsules are the components of passengers or cargo. Capsules are the car body support mechanism by the beam, column, floor beam, and oblique rods, etc. It has load base body, enclosure, car roof ventilation and lighting, car decoration and capsules of inner control button, etc. The car body space size is determined by the rated load and rated number of guests.

Fifth, the weight balance system

Weight balance system is composed of compensation device of weight of heavy rack and the heavy blocks. For heavy part will balance the weight of capsules and the rated load. Weight compensation device is compensate high-rise elevator capsules and lead on the heavy side pulling wire rope length on the impact of change on elevator balance.

Basic PLC and Elevator Control System

Sixth, the electric drive system

Electric drive system is built on pulling motor, power supply system, speed feedback device, control device and so on; its action is carried out on the elevator speed control. Pulling motor is the power source of the elevator. According to the elevator configuration, we can use ac motor or dc motor. The motor power supply system is to provide power to device. Speed feedback system is to provide elevator speed signal for speed regulating system. Generally, velocity pulse generator is connected to the motor. Speed regulating device is pulling motor speed control.

Seven, electrical control system

The elevator electric control system is used to control unit, control devices, flat layer device, and the location of display device and other parts. According to the logic of the elevator function requirements, the control device uses to control the operation of the elevator which is set in the computer room controlling on the ark. Controlling device is composed of capsules' inside button and hall door call box button to manipulate the running of the lift. Flat layer device is uses to control flat layer signals. Flat layer, refers works in a special way to close the door when it leaves a certain floor. Display device is used to display the elevator floor location inside the capsules and the light on the hall door, with indicating the direction of the lift.

Eighth, security system

Safety protection system includes the protection of the mechanical system and electrical protection system, used to protect the safety of the elevator. Mechanical aspects are: the speed limiter and safe pliers over-speed protection, buffer header and hit bottom protection, and cut off the limit of the total power supply protection device. Electrical safety protection is evident in every aspect of running the lift.

1.2.4 The Main Parameters and Size of an Elevator

1.2.4.1 The main parameters of the elevator:

1. Rated load (kg): manufacturing and design requirements of the elevator load.

2. Capsules dimensions (mm): height × width × depth.

3. The car forms: single or double door switch and other special requirements such as the car roof, car bottom, handle, color choices.

4. Door sedan forms: Three fence gate, carved doors closed, closed the double doors demolition, demolition carved double doors closed.

Md. Lutfar Rahman

5. Door width (mm): Net width of the capsules door and landing door fully open.

6. Opening direction: people in the hall facing the hall door, the door to the left direction will open in left, right door in right. The door will open left and right respectively.

7. The traction mode: The speed of the car is equal to the speed of the wire.

8. Rated speed (m / s): Manufacturing design specified speed and lift.

9. Electrical control system: It includes drag system such as AC or DC motor drag and button of inside and outside of the capsules.

10. Stop Layer Station: Each floor within the building is used to access the location of the capsules are referred to as the station.

11. Lifting height (mm): It is underlying from the end station floor to the top layer of the vertical distance.

12. Top-level height (mm): It refers to the insulation layer from the bottom of the room to the top end of the vertical distance.

13. Pit depth (mm): It refers to the bottom floor to the vertical distance between the undersides of the shaft of the elevator, the faster the speed, the more deep pit in general.

14. The well depth (mm): For the sound insulation the well depth should be the vertical distance between the shaft floor and the building component.

1.2.4.2 The size for the standard elevator:

In order to strengthen and improve of the elevator products, there is some standard form established such as JBI816 JBI435-74-74, JB/Z110-74 in 1974. Main parameters of the lift is the basis of design and manufacture of the elevator .When we install an elevator we should focus the location, carrying objects, the correct selection of the elevator and its related parameters and dimensions. According to these parameters and the sizes, design and construction should be made in the specified building. Otherwise it will cause the bad effect of the elevator.

Chapter 2 Request of the Elevator Control System

2.1 Control Requirements

Requirements for going up direction:

i. When the elevator is on 1st or 2nd 3rd layer, if someone presses the SB4 button then the elevator goes up and stops 4th floor.
ii. When the elevator is on 1st floor and SB2 button are pushed then it will rise up and stop 2nd floor, if pressed SB3 it will reach on 3rd floor and then stop.
iii. When the elevator is on 2nd floor and SB3 button are pushed then it will rise and stop 3rd floor.
iv. When the elevator is in 1st floor and SB2 and SB3 button are pushed simultaneously, the elevator will stop on 2nd floor for the suspension then continue to go up for 3rd floor.
v. Staying elevator on 2nd floor, SB3 and SB4 button are pushed for going up call then it will first stop 3rd floor to let people leave, then it will go up and stop 4th floor.
vi. When the elevator stops at 1st floor, SB2 and SB4 button are pushed for call up then it will first stop on 2nd floor to let people leave and then it will go up and stop on 4th floor.
vii. When the elevator stop on 1st floor, SB3 and SB4 button are pushed, then it will first stop on 3rd floor to let people leave and then it will continue to go up and stop 4th floor.
viii. When the elevator stops on 1st floor, SB2,SB3 and SB4 button are pushed then it will first stop 2nd and then 3rd floor to let people leave and finally it will arrive at 4th floor.

Requirements for going down direction:

i. When the elevator is on 4th or 3rd or 2nd floor and SB1 button are pushed then elevator will go down and stop at 1st floor.
ii. When the elevator is on 4th floor and SB3 button is pushed and elevator will descend on 3rd floor, if SB2 button is pushed it will continue to go down and stop on 2nd floor.
iii. When the elevator is on the 3rd floor and SB2 button are pushed and the elevator will stop on 2nd floor.
iv. When the SB3 and SB2 button are pushed at the same time while staying on 4th floor, it will stop 3rd then stop at 2nd floor.
v. When the elevator is on 4th floor and SB3 and SB1 button are pressed at the same time then it will stop 3rd floor then 1st floor.

Basic PLC and Elevator Control System

vi. When the elevator is on 4th floor and SB3, SB2 and SB1 buttons are pushed at the same time, it goes down and stops on 3rd, 2nd and 1st floor respectively.

vii. When the elevator is on 3rd floor and SB2 and SB1 buttons are pushed at the same time then it goes down and stops 2nd and 1st floor respectively.

viii. When the elevator is on 2nd floor and SB1 button is pushed and the elevator will go down and stop at 1st floor.

<u>Requirements for going up in Critical Condition:</u>

i. The elevator is in 1st floor and SB3, SB6, SB4 buttons are pushed, the elevator will not stop on 2nd floor; first it will stop on 3rd floor to let the people leave after thon it will reach on 4th floor for suspension. While goes down the elevator will stop on 2nd floor to let the people come in.

ii. The elevator is on the 2nd floor and SB1, SB10 button are pushed then the elevator will go up and stop on the 4th floor and let the people come in and then goes down to the 1st floor.

iii. The elevator is on the 2nd floor and SB8, SB5 button are pushed on the same time then the elevator will go up and stop on the 3rd floor to let people come in and then goes down to the 1st floor.

iv. The elevator is on the 3rd floor and SB8, SB5 button are pushed on the same time then the elevator will go up and stop on the 3rd floor and let the people come in and then goes down to the 1st floor.

2.2 Process Solutions

The main component of electric part of the elevator system is the motor drive signal and internal and external control button of the elevator. To set up and use of these resources rationally, is very important matter to solve the problem. First, we have to understand the characteristics of the controlled object to design then have to make sure about the number of input and output points of PLC. For a four layers of elevator control, to solve the main problems we have to focus the following aspects:

i. Settings of the Elevator and Hall button carefully.
ii. Monitoring the location of elevator operation.
iii. Monitoring the open and close state of the elevator and the hall door carefully.
iv. To determine the control logic.

By analyzing the above questions, we can complete the requirements by choosing the PLC. In logic-based control systems, we focus to control of motor forward and reverse direction. When the elevator or the hall buttons are pressed to ascend or descend, the motor should be controlled according to the movement of the elevator. When a call should be made or arrived the elevator responses the priority signal in running direction and then response to another direction.

Chapter 3 Hardware System Configuration

3.1 Elevator System Design

PLC has the different type with different functions and commands, but the structure and working principle is similar. Usually, PLC consists of several main part of the host input / output interface, power supply, programming interface and external expansion device and other components. According to the requirements of the elevator control system, I chose Chinese Siemens Company launched small PLC S7-200 series; it has high reliability, small volume, convenient extension, and the advantages of using flexibility. For more input and output I choose CPU 226 series with EM222 extension module.

3.2 PLC Selection

3.2.1 Overview of Siemens S7-200 PLC

Siemens S7-200 PLC system is a compact, programmable logic controller. It has the good hardware architecture system, friendly CPU module and extensive expansion modules. It can satisfy the needs of automation control of various devices. In addition, S7-200 PLC has basic control functions, but also has unique following aspects.

    i.   A powerful instruction set

Instruction includes the bit logic instructions, counters, timers, complex numbers arithmetic instructions, string instructions, clock instructions, as well as dedicated and intelligent modules.

    ii.   The rich and powerful communications capabilities

S7-200 offers nearly 10 kinds of communication to satisfy different application requirements. For the simple communication between the Profibus-DP network communications to the S7-200 we can also use Ethernet communications. We can say that a communication function of S7-200 has far exceeded the level of overall communication compact in PLC.

    iii.   The simple use of Programming Software

Step 7 - Micro / WIN V9.0 programming software provides users with the development, editing, and good programming environment.

    iv.   Constant Innovation

Coinstantaneous innovation of Siemens Company is common principle that is reflected most vividly. The company never stops launching the new products which makes them to become the market leader in S7-200.

3.2.2 PLC and its Application Characteristics of the Elevator Control

3.2.2.1 PLC Features:

PLC is a kind of special computer for industrial automation control, essentially belongs to the computer control mode. Like ordinary microcomputer, PLC controllers is to make general or special word processor, the CPU implementation channels (words) like arithmetic and data storage, and a processor (Boolean processor), point (a) operation and control. PLC controls with high reliability, easy operation, simple maintenance, programming, flexibility, etc.

Reliability: Serviceable products, including the product's effectiveness and reliability.

i. PLC does not require a lot of activity and wiring connection. So the wiring is greatly reduced. At the same time, the system is simple to maintenance, and repair time is short.
ii. PLC uses a series of reliability design method for the design, such as redundant design, power protection, fault diagnostics and data protection and recovery. It also improves MTBF, MTTR reduced and thus the reliability is improved.
iii. In terms of hardware, PLC is using a series of measures to improve the reliability. For example, the use of components reliability, advanced technology, manufacturing assemble line, interference shielding, isolation and filtering, against power outage protection and protection of memory contents, etc.
iv. In terms of software, PLC has also taken a series of measures to improve the reliability of the system. For example, the use of filtering software, software self-diagnostics and simplify programming language.

Easy operation: operability performance of PLC is the following several aspects,

i. PLC has convenient operation, including programming input and the change of operating procedures. Most PLC programming operation uses the input and changes. Programmer input information to provide at least a display of medium-sized PLC, the programmer uses a display, so that we can directly enter the program. Changing the program's operation can be performed directly from a search or in a order to find the desired address number or the contact number, and then we can make changes.
ii. The variety of programming language is available. For electrical technicians, due to the ladder diagram and electrical schematic diagram relatively close, it is easy to learn and understand.

iii. Convenient maintenance of PLC has self-diagnosis function requirement. When a system failure happens, it does diagnoses through hardware and software maintenance.

Flexibility: The flexibility of the PLC is in the following areas,

There are several programming languages for PLC including ladder diagram, SFC, STL, ST and so on. If operator can master only one of programming languages, he can operate PLC well. Anyone who wants to use PLC has a good choice. Based on engineering practice, capacity and function can be expanded by expanding number of module, so PLC has a good flexibility.

Strong in Quality and Easy-Operating: It is very easy to edit and modify program for PLC by computer offline or online. It is very easy to find out where the fault lies by observing the information of fault and function of self diagnosing. Its function to maintain and repair is so much easy. It is very easy to configure PLC because of the modularization, standardization, serialization of PLC.

3.3 The Idea of PLC Control System Design

According to the working process and requirements of the elevator design, Siemens S7-200 PLC acts as the core, which works for the process to display and control. PLC control system design in general is divided into the following steps:

i. We should be familiar with the controlled object to develop control programs.
ii. We should make sure about I/O status.
iii. Select the PLC model.
iv. Select the input and output devices, the distribution of the PLC I / O address.
v. Programming design (including the ladder drawing).
vi. System debugging.

Basic PLC and Elevator Control System

Overall system design as shown in the following figure:

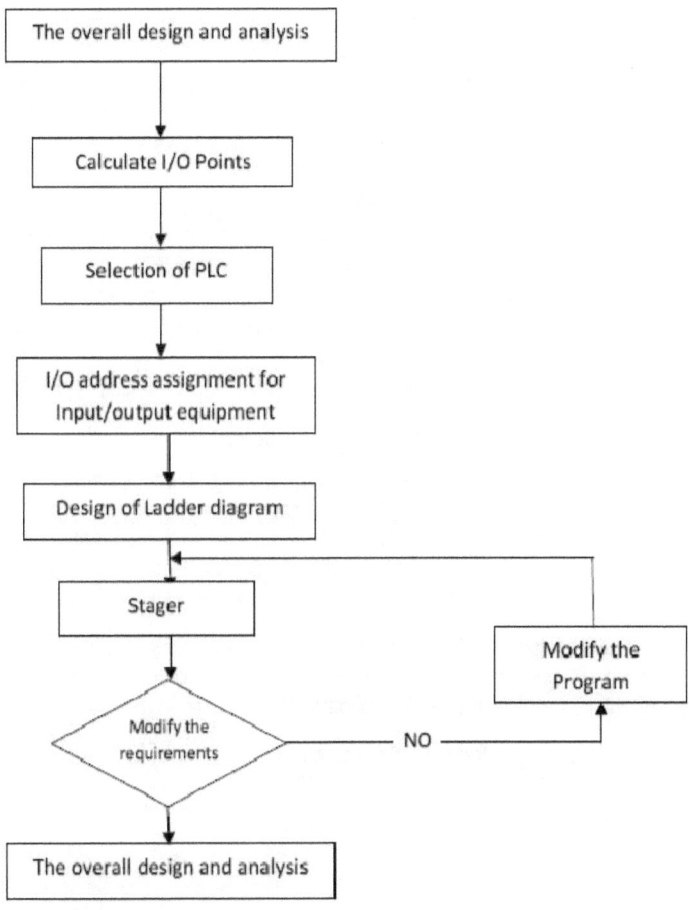

Figure 3.1: Overall system design of elevator control

3.4 The Main Research Content

This design is the use of a programmable logic controller (PLC) to control a four-layer elevator. The first hardware design includes the CPU processing module, expansion module, resource allocation, and external wiring, etc. And, then on the basis of hardware design, the choice of Siemens S7-200 PLC system design process flow diagram and ladder program, after that the system is analyzed and debugging to achieve automatic control of the elevator. For Electrical Control Design,

3.4.1 PLC Input / Output Point Estimation

To constitute four layer elevator electrical control systems, we use motor to elevator up and down. We have used travel switch SQ1~ SQ4, button SB1~SB10. For the light indication we have used E1~E10, to move up and down of the elevator we use KM1 and KM2. The input includes SB1~SB10, open door and close door as well as travel switch SQ1~ SQ4. The output includes E1~E10, KM1 and KM2 as well as make door open and close button. A total of 16 input points and total of 14 that makes a total of 30 points.

3.4.2 The Estimation of PLC Memory Capacity

User control procedures requires memory capacity and memory utilization, input / output points, programming level and other factors. Therefore, before programming can only be based on the user input / output points, and then the complexity of the control system is estimated. The system has a switch I / O has 30 total points, analog I / O number are 0. PLC memory use formulas to estimate the total capacity:

Deposit words = switch I / O total points × (10 ~ 15) + Analog I / O total points × (150 ~ 250) and then set aside about 30% margin. The system needs to estimate approximately 1K bytes of memory capacity.

In consideration of Integrated I / O points, and memory capacity, total input and output points (I/O points 24/16), we have used EM222 module as an expansion with CPU226 that can easily meet the requirements.

Basic PLC and Elevator Control System

3.5 PLC Input / Output Assignment

The system takes the 30 input and ports, 16 inputs, 14 outputs, specific I / O assignment as shown in Table 3-5-1 and Table 3-5-2:

Sequence	Input interface(I)	Description	Symbol
1	I 0.1	"1st floor call" **inside** elevator	SB1
2	I 0.2	"2nd floor call" **inside** elevator	SB2
3	I 0.3	"3rd floor call" **inside** elevator	SB3
4	I 0.4	"4th floor call" **inside** elevator	SB4
5	I 0.5	"1st floor call" **outside** elevator	SB5
6	I 0.6	"2nd floor down call" in **outside**	SB6
7	I 0.7	"2nd floor up call" in **outside**	SB7
8	I 1.0	"3rd floor down call" in **outside**	SB8
9	I 1.1	"3rd floor up call" in **outside**	SB9
10	I 1.2	"4th floor call" **outside** elevator	SB10
11	I 1.3	Travel switch in 1st floor	SQ1
12	I 1.4	Travel switch in 2nd floor	SQ2
13	I 1.5	Travel switch in 3rd floor	SQ3
14	I 1.6	Travel switch in 4th floor	SQ4
15	I 1.7	Call for open door of the elevator	n/a
16	I 2.0	Call for close door of the elevator	n/a

Table 3-5-1: PLC input assignment:

Basic PLC and Elevator Control System

Sequence	Output Interface(Q)	Description	Symbol
1	Q 0.1	Light of "1st floor"	E1
2	Q 0.2	Light of "2nd floor"	E2
3	Q 0.3	Light of "3rd floor"	E3
4	Q 0.4	Light of "4th floor"	E4
5	Q 0.5	Call light of "1st floor"	E5
6	Q 0.6	Call down light of "2nd floor"	E6
7	Q 0.7	Call up light of "2nd floor"	E7
8	Q 1.0	Call down light of "3rd floor"	E8
9	Q 1.1	Call up light of "3rd floor"	E9
10	Q 1.2	Call light of "4th floor"	E10
11	Q 1.3	Elevator comes down	KM1
12	Q 1.4	Elevator goes up	KM2
13	Q 1.5	Make door open	n/a
14	Q 1.6	Make door close	n/a

Table 3-5-2: PLC output assignment:

3.6 PLC Control Circuit Design

According to the requirement of overall plan and the PLC I/O port distribution table, the circuit schematic diagram is shown to the next page:

Basic PLC and Elevator Control System

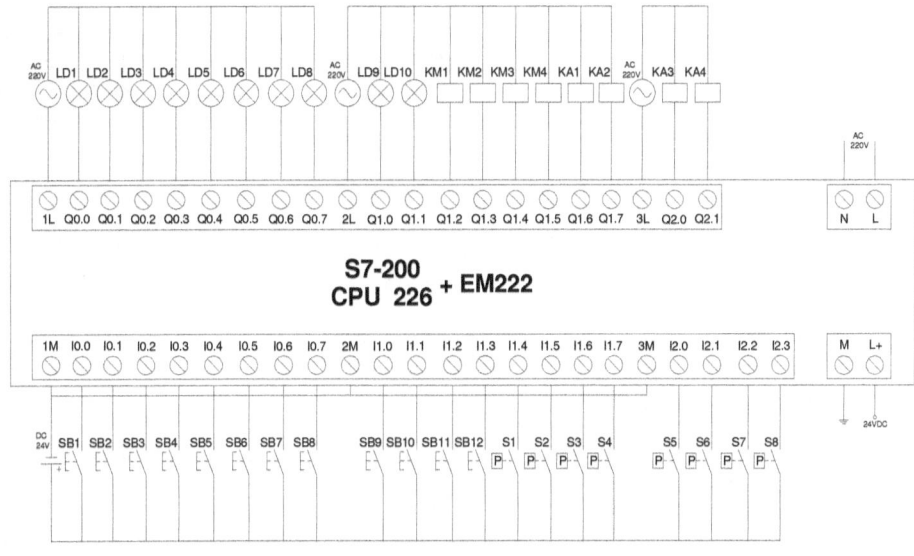

Figure 3.2: Principle Diagram of Control Circuit.

3.7 Main Circuit Diagram

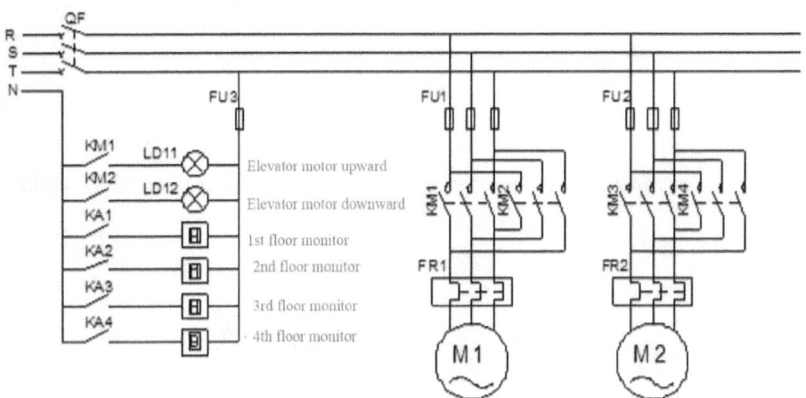

Chapter 4 Software Design

STEP7-Micro/WIN is designed specifically for the S7-200, under the personal computers it is running the Windows operating system. It is powerful, easy to use, and easy to learn. Users can make Statement List, Ladder Diagram and Function Block Diagram programming. The program can be prepared by different programming languages also can be converted to each other. We can use the symbol table to define the variables used in the program address corresponding symbol.

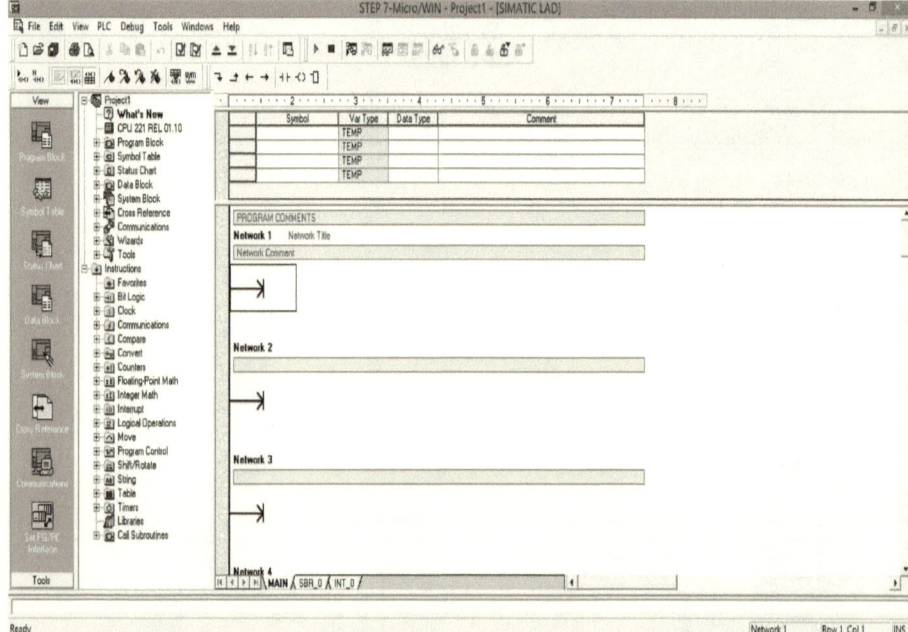

Figure 4.1 program editor window for STEP7 - Micro/WIN4.0 programming interface.

4.1 Program Design

4.1.1 Program Flowchart

Design Elevator PLC controlled automatic operation mode, the control logic design based on this flow chart shown in Figure 4.2

Basic PLC and Elevator Control System

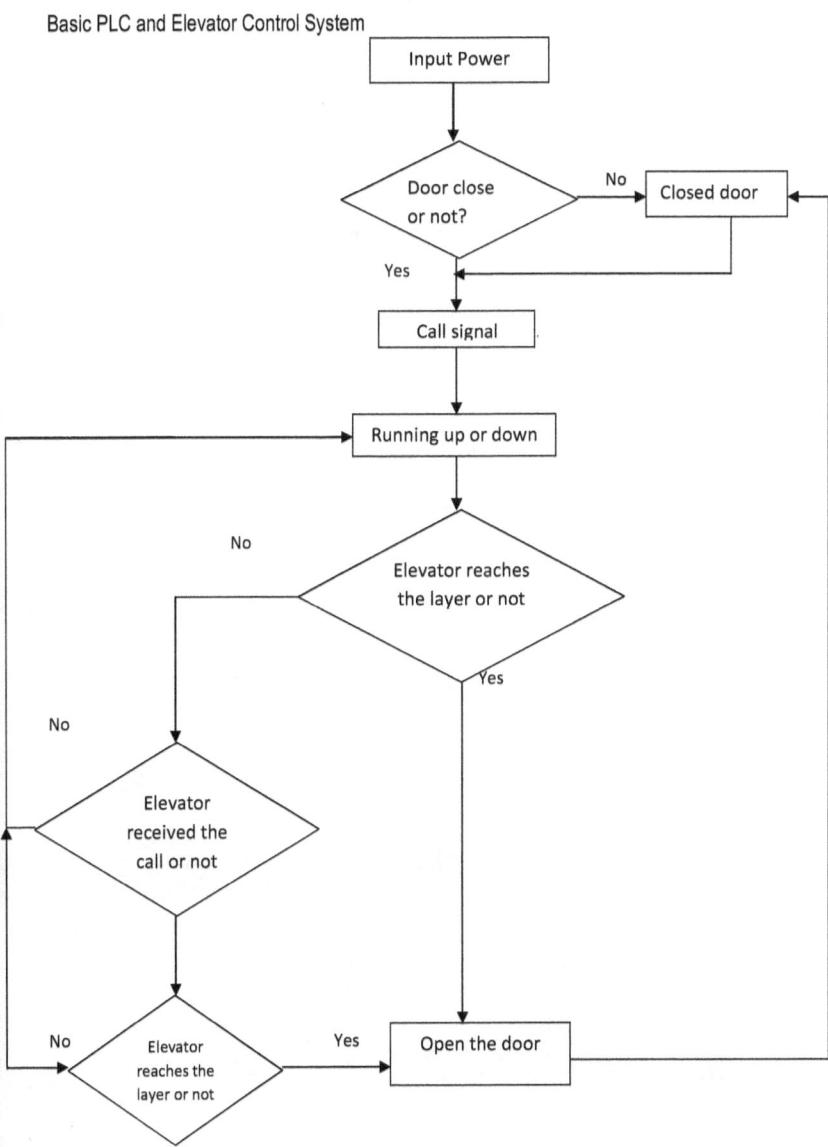

Figure 4.2: Flow chart of control logic of elevator

Basic PLC and Elevator Control System

4.1.2 Flowchart of the Elevator Up and Down

The lift up and down flow chart shown in figure 4.3

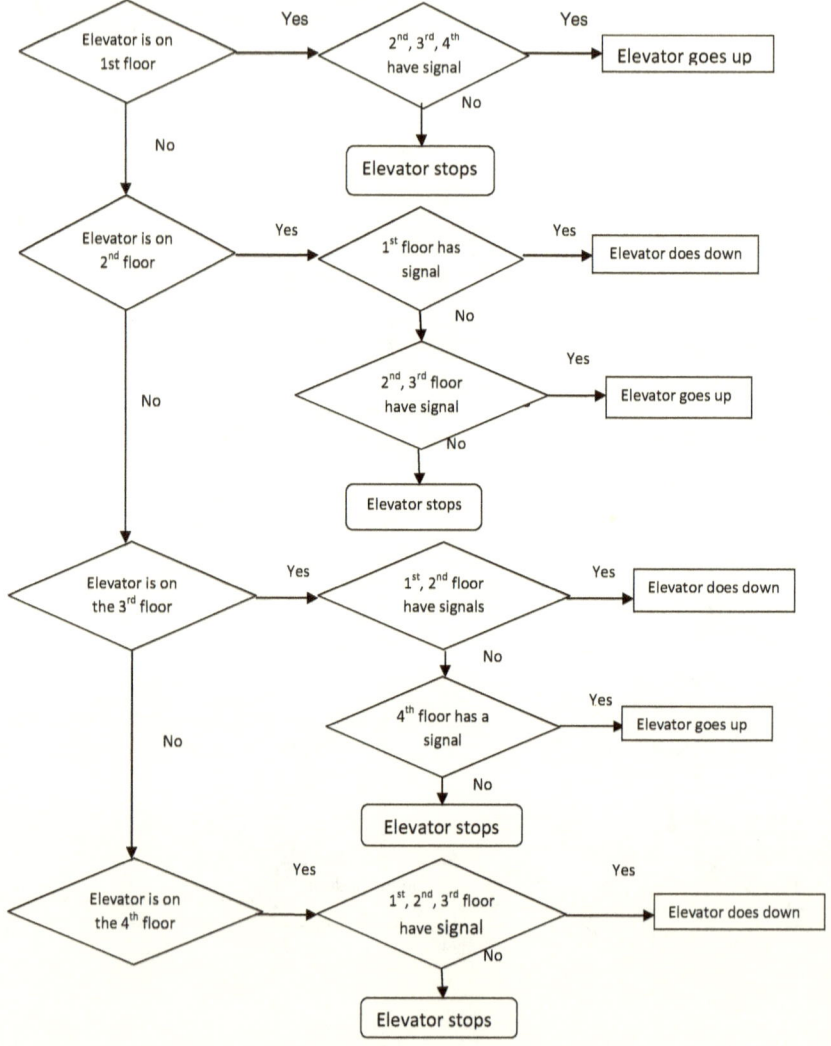

Figure 4.3: The flow chart of the elevator up and down.

Basic PLC and Elevator Control System

4.1.3 Ladder Logic Diagram:

Network 1:

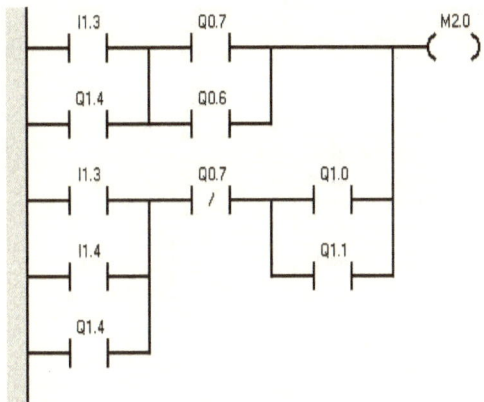

Symbol of Network 1	Address
Call down light of "2nd floor"	Q0.6
Call down light of "3rd floor"	Q1.0
Call up light of "2nd floor"	Q0.7
Call up light of "3rd floor"	Q1.1
Elevator goes up	Q1.4
Travel switch in 1st floor	I1.3
Travel switch in 2nd floor	I1.4

Network 2:

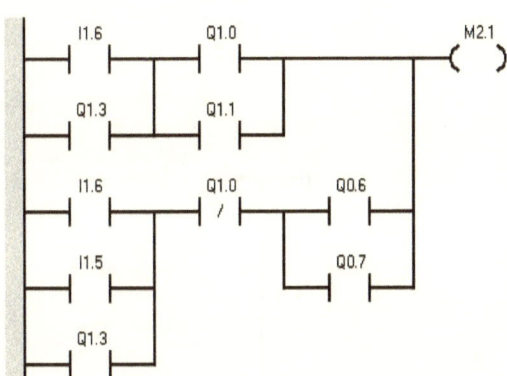

Symbol of Network 2	Address
Call down light of "2nd floor"	Q0.6
Call down light of "3rd floor"	Q1.0
Call up light of "2nd floor"	Q0.7
Call up light of "3rd floor"	Q1.1
Travel switch in 1st floor	Q1.3
Travel switch in 1st floor	I1.5
Travel switch in 4th floor	I1.6

Basic PLC and Elevator Control System

Network 3:

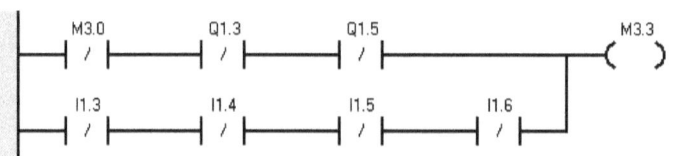

Symbol of Network 3	Address
Elevator comes down	Q1.3
Elevator goes up	Q1.5
Travel switch in 1st floor	I1.3
Travel switch in 2nd floor	I1.4
Travel switch in 3rd floor	I1.5
Travel switch in 4th floor	I1.6

Symbol of Network 4	Address
Elevator goes up	Q1.4
Elevator goes up	Q1.5
Travel switch in 1st floor	I1.3
Travel switch in 2nd floor	I1.4
Travel switch in 3rd floor	I1.5
Travel switch in 4th floor	I1.6

Network 4:

Network 5:

Md. Lutfar Rahman

Basic PLC and Elevator Control System

Symbol of Network 5	Address
Call light of "4th floor"	Q1.2
Call up light of "2nd floor"	Q0.7
Call up light of "3rd floor"	Q1.1
Elevator goes up	Q1.4
Travel switch in 1st floor	I1.3
Travel switch in 2nd floor	I1.4
Travel switch in 3rd floor	I1.5

Symbol of Network 6	Address
Call down light of "2nd floor"	Q0.6
Call down light of "3rd floor"	Q1.0
Call light of "1st floor"	Q0.5
Elevator comes down	Q1.3
Travel switch in 2nd floor	I1.4
Travel switch in 3rd floor	I1.5
Travel switch in 4th floor	I1.6

Network 6:

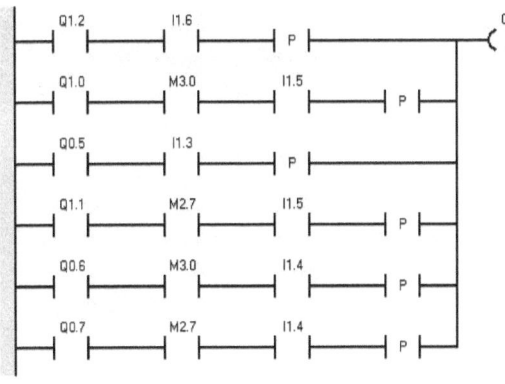

Network: 7

Symbol of Network 7	Address
Call down light of "2nd floor"	Q0.6
Call down light of "3rd floor"	Q1.0
Call light of "1st floor"	Q0.5
Call light of "4th floor"	Q1.2
Call up light of "2nd floor"	Q0.7
Call up light of "3rd floor"	Q1.1
Make door open	Q1.5
Travel switch in 1st floor	I1.3
Travel switch in 2nd floor	I1.4
Travel switch in 3rd floor	I1.5
Travel switch in 4th floor	I1.6

Basic PLC and Elevator Control System

Network: 8

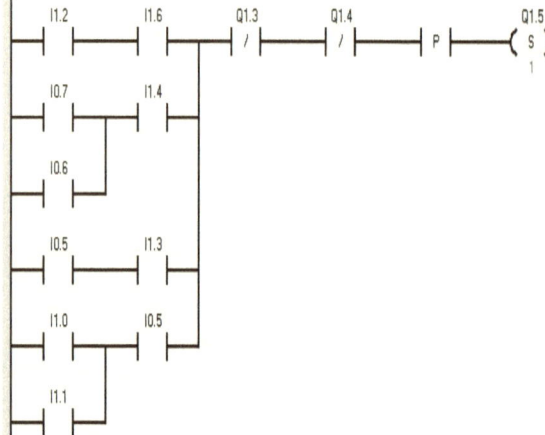

Symbol of Network 8	Address
"2nd floor down call" in outside	I0.6
"3rd floor down call" in outside	I1.0
"1st floor call" outside elevator	I0.5
"4th floor call" outside elevator	I1.2
"2nd floor up call" in outside	I0.7
"3rd floor up call" in outside	I1.1
Elevator comes down	Q1.3
Elevator goes up	Q1.4
Make door open	Q1.5
Travel switch in 1st floor	I1.3
Travel switch in 2nd floor	I1.4
Travel switch in 4th floor	I1.6

Network: 9

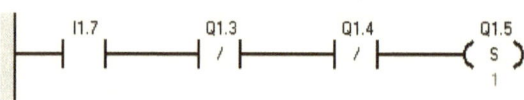

Symbol of Network 9	Address
Elevator comes down	Q1.3
Elevator goes up	Q1.4
Make door open	Q1.5
Call for open door	I1.7

Network 10:

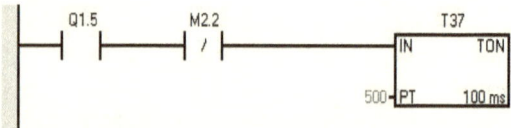

Symbol of Network 10	Address
Make door open	Q1.5

Md. Lutfar Rahman

Network 11:

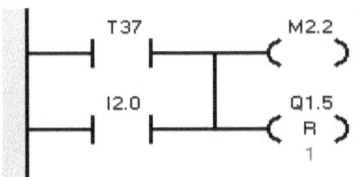

Symbol of Network 11	Address
Call for close door	I2.0
Make door open	Q1.5

Network 12:

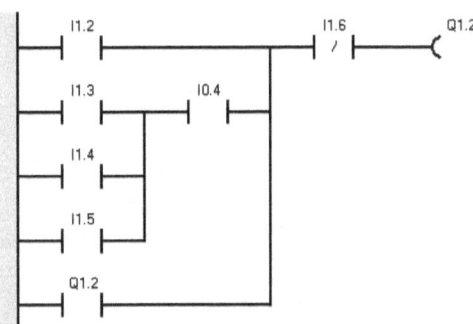

Symbol of Network 12	Address
"4th floor call" inside elevator	I0.4
"4th floor call" outside elevator	I1.2
Call light of "4th floor"	Q1.2
Travel switch in 1st floor	I1.3
Travel switch in 2nd floor	I1.4
Travel switch in 3rd floor	I1.5
Travel switch in 4th floor	I1.6

Network 13:

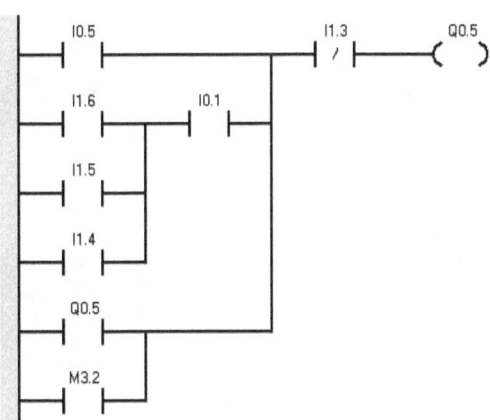

Symbol of Network 13	Address
"1st floor call" inside elevator	I0.1
"1st floor call" outside elevator	I0.5
Call light of "1st floor"	Q0.5
Travel switch in 1st floor	I1.3
Travel switch in 2nd floor	I.14
Travel switch in 3rd floor	I1.5
Travel switch in 4th floor	I1.6

Basic PLC and Elevator Control System

Network 14:

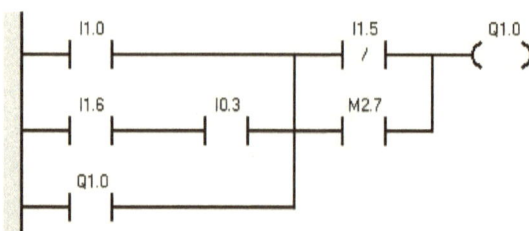

Symbol of Network 14	Addr
"3rd floor call" inside elevator	I0.3
"3rd floor down call" in outside	I1.0
Call down light of "3rd floor"	Q1.0
Travel switch in 3rd floor	I1.5
Travel switch in 4th floor	I1.6

Network 15:

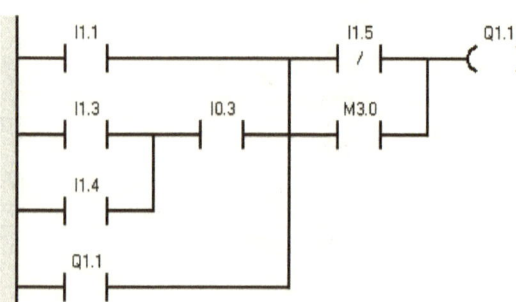

Symbol of Network 15	Addr
"3rd floor call" inside elevator	I0.3
"3rd floor up call" in outside	I1.1
Call up light of "3rd floor"	Q1.1
Travel switch in 1st floor	I1.3
Travel switch in 2nd floor	I1.4
Travel switch in 3rd floor	I1.5

Network 16:

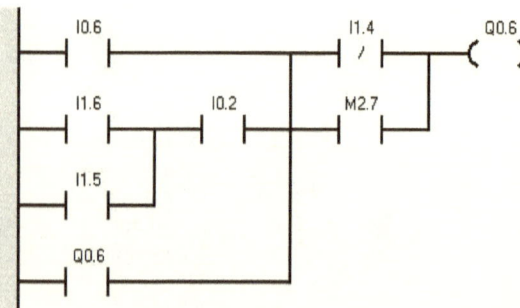

Symbol of Network 16	Addr
"2nd floor call" inside elevator	I0.2
"2nd floor down call" in outside	I0.6
Call down light of "2nd floor"	Q0.6
Travel switch in 2nd floor	I1.4
Travel switch in 3rd floor	I1.5
Travel switch in 4th floor	I1.6

Md. Lutfar Rahman

Basic PLC and Elevator Control System

Network 17:

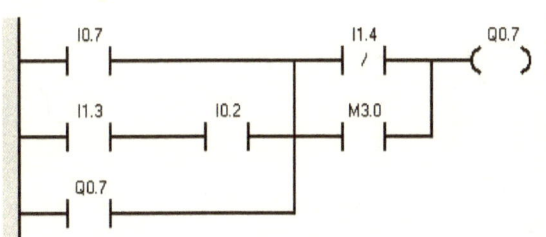

Symbol of Network 17	Address
"2nd floor call" inside elevator	I0.2
"2nd floor up call" in outside	I0.7
Call up light of "2nd floor"	Q0.7
Travel switch in 1st floor	I1.3
Travel switch in 2nd floor	I1.4

Network 18:

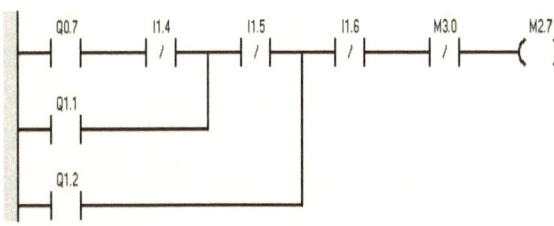

Symbol of Network 18	Address
Travel switch in 2nd floor	Q1.2
Call up light of "2nd floor"	Q0.7
Call up light of "3rd floor"	Q1.1
Travel switch in 2nd floor	I1.4
Travel switch in 3rd floor	I1.5
Travel switch in 4th floor	I1.6

Network 19:

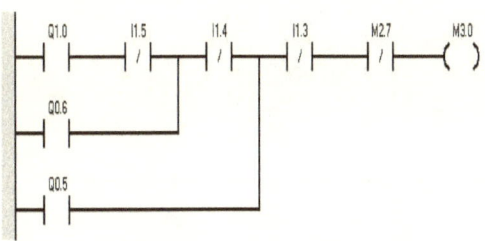

Symbol of Network 19	Address
Call down light of "2nd floor"	Q0.6
Call down light of "3rd floor"	Q1.0
Call light of "1st floor"	Q0.5
Travel switch in 1st floor	I1.3
Travel switch in 2nd floor	I1.4
Travel switch in 3rd floor	I1.5

Basic PLC and Elevator Control System

Network: 20

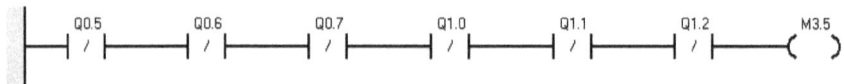

Symbol of Network 20	Address
Call down light of "2nd floor"	Q0.6
Call down light of "3rd floor"	Q1.0
Call light of "1st floor"	Q0.5
Call light of "4th floor"	Q1.2
Call up light of "2nd floor"	Q0.7
Call up light of "3rd floor"	Q1.1

Symbol of Network 21	Address
Make door open	Q1.5
Travel switch in 1st floor	I1.3

Network 21:

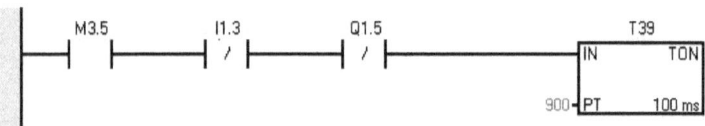

Network 22:

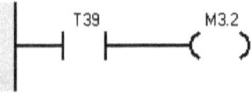

Symbol of Network 23	Address
Travel switch in 1st floor	I1.3
Travel switch in 2nd floor	I1.4
Travel switch in 3rd floor	I1.5
Travel switch in 4th floor	I1.6

Network 23:

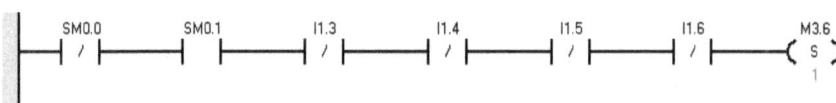

Md. Lutfar Rahman

Basic PLC and Elevator Control System

Network 24:

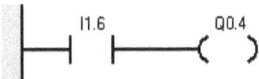

Symbol of Network 24	Address
Light of "4th floor"	Q0.4
Travel switch in 4th floor	I1.6

Network 25:

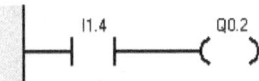

Symbol of Network 25	Address
Light of "2nd floor"	Q0.2
Travel switch in 2nd floor	I1.4

Network 26:

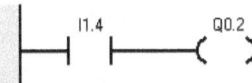

Symbol of Network 26	Address
Light of "2nd floor"	Q0.2
Travel switch in 2nd floor	I1.4

Network 27:

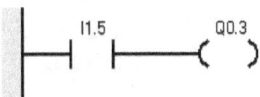

Symbol of Network 27	Address
Light of "3rd floor"	Q0.3
Travel switch in 3rd floor	I1.5

Network 28:

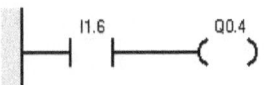

Symbol of Network 28	Address
Light of "4th floor"	Q0.4
Travel switch in 4th floor	I1.6

4.2 Program Debugging

After completing the hardware design, production and software programming, we have to debug the entire system to run normal operation. System debugging includes hardware debugging and system debugging software debugging.

4.2.1 The Hardware Debugging

Like the normal electrical wiring PLC practical wiring also should consider the following aspects:

i. It should have the power input line, usually for 220 v, 50 hz ac power, allows for a certain range to the power supply. It must have a protection device, such as fuse, etc. If strong interference or high reliability requirements need in any occasion, it should be retrofitted with a shield at the power input of the PLC isolation transformer and low-pass filter.
ii. For the group of eight input terminals, public COM should be connected with separate ground and other electrical components do not share the grounding wire, grounding wire should be larger than input terminals for a group of eight, pay a COM port. PLC should separate grounding, don't and other electrical components Shared ground wire, ground wire squared area should be greater than 2 mm^2 as close to the plc.
iii. PLC outputs have a coil and the perceptual components such as solenoid valve must add to protection circuit

4.3 Software Debugging

The user can use the programming tools in the computer to make program, after repeated edit, compile, download, debug and run until run correctly.

4.4 Edit and Compile

Ladder Editor opens the program in the computer. After the program is entered, using the pull-down menu or toolbar shortcut button to compile the program, the compilation output results can be displayed in the input window below the display. If anything goes error we can clearly point out the wrong network segment, then we can edit and correct it after that we can compile until compile correctly.

4.5 Download

After the program compiles successfully, by clicking the download shortcut button we can download the selected content to the PLC memory.

Basic PLC and Elevator Control System

4.6 Monitoring Program and Running the Debug

Program editor also can be used to monitor the program execution. To see the each components status we can open the process of data by opening debug menu. Then it will show the contact and electrified coil internal color blue. In the operation of the PLC working, status changes with input conditions, timing, counting and the operation of the process. Each scan cycle of input phase refreshes according to period of time. The online dynamic observation program is running so we find the errors easily and then can change it. After all aspects of the system debugging, the system run correctly and show us the design is reasonable.

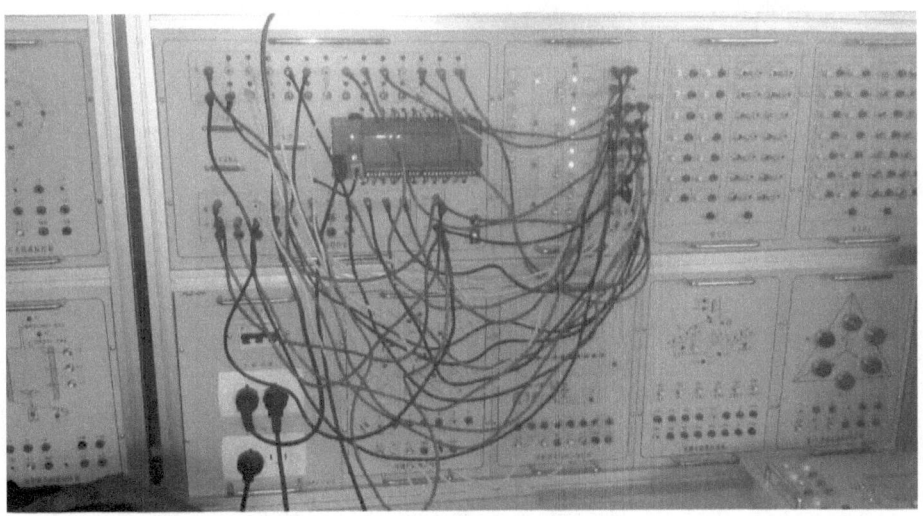

Figure 4.4: Monitoring the program in real life

Chapter 5 Conclusion

5.1 Check

I simulate my PLC program of four levels Elevator in two ways. First I simulate it in simulator software and secondly I simulate it in hardware. In the both case, my program runs successfully and I check it from my project teacher. Although some calibrations and requirements may have, the modeling PLC based on elevator control system is done. The traditionally used relays and IC boards have been replaced by PLC for easy and cheap controlling mechanism used in this elevator. By developing this system, the result of elevator control system can be applied in the real world. By using PLC based elevator control system, the desired position can be forecasted. The simulation results of the four floors system have been discussed. As a future work, my PLC based elevator model is intended to construct and tested to be applicable in the real world.

5.2 Summary

The design is basically achieved the design goal. To achieve the goal, I have to use common PLC control which has reasonable equipment selection, parameter setting and software designed to improve the reliability of the operation of the elevator. In the elevator control system, elevator should run comfortably and must save the energy. Through this graduation project, I am not only benefited to master my in this topics but also broadened my knowledge. I also recognized more clearly some of the shortcomings. Because of my limited ability, this design has some shortcomings, such as: The design is plotted for four, no consideration of the elevator door motor control and fire-fighting operation and so on. Besides, in the distribution of hardware devices are connected between the distributions I / O port address, these areas did not very understand at first, but after six months of self-study, as well as the help of my respectable teacher teachers and beloved classmates I have gained a depth understanding. Through six months of practice and learning, I learned a lot of knowledge that does not only relate to the textbook but also field experience. It has given me the feelings of joy and success. I will continue to improve my knowledge for further studies to strive the greater achievements.

Md. Lutfar Rahman

Basic PLC and Elevator Control System

Appendix A – About S7-200

Understanding How the S7-200 Executes Control Logic

The S7-200 continuously cycles through the control logic in program, reading and writing data.

i. The S7-200 Relates Program to the Physical Inputs and Outputs

The basic operation of the S7-200 is very simple:

- The S7-200 reads the status of the inputs.

- The program that is stored in the S7-200 uses these inputs to evaluate the control logic. As the program runs, the S7-200 updates the data.

- The S7-200 writes the data to the outputs.

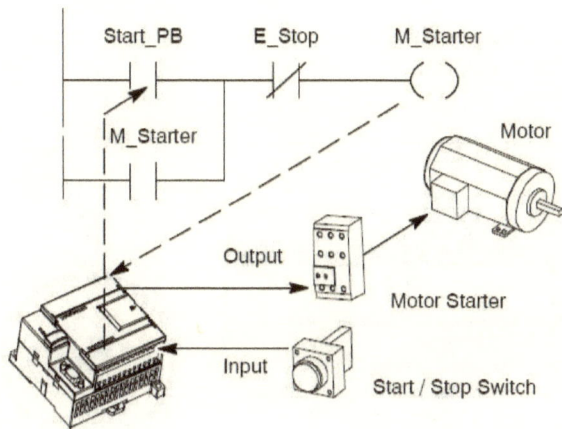

Figure Appendix A.1: Controlling Inputs and Outputs

Figure 4-1 shows a simple diagram of how an electrical relay diagram relates to the S7-200. In this example, the state of the switch for starting the motor is combined with the states of other inputs. The calculations of these states then determine the state for the output that goes to the actuator which starts the motor.

Basic PLC and Elevator Control System

ii. The S7-200 Executes Its Tasks in a Scan Cycle

The S7-200 executes a series of tasks repetitively. This cyclical execution of tasks is called the scan cycle. As shown in Figure 4-2, the S7-200 performs most or all of the following tasks during a scan cycle:

- Reading the inputs: The S7-200 copies the state of the physical inputs to the process-image input register.

- Executing the control logic in the program: The S7-200 executes the instructions of the program and stores the values in the various memory areas.

- Processing any communications requests: The S7-200 performs any tasks required for communications.

- Executing the CPU self-test diagnostics: The S7-200 ensures that the firmware, the program memory, and any expansion modules are working properly.

- Writing to the outputs: The values stored in the process-image output register are written to the physical outputs.

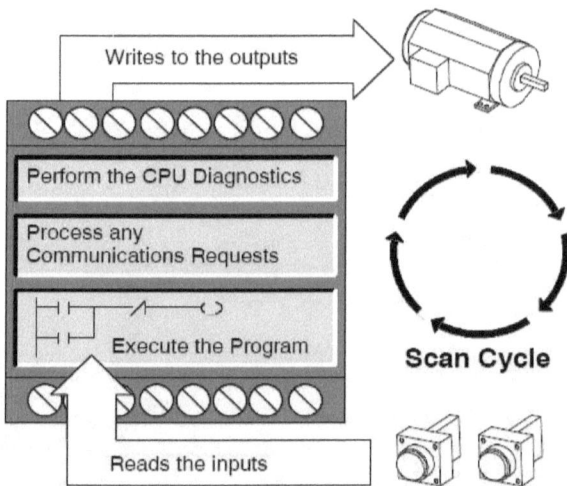

Figure Appendix A.2: S7-200 Scan Cycle

The execution of the user program is dependent upon whether the S7-200 is in STOP mode or in RUN mode. In RUN mode, program is executed; in STOP mode, program is not executed.

Md. Lutfar Rahman

iii. Reading the Inputs

Digital inputs: Each scan cycle begins by reading the current value of the digital inputs and then writing these values to the process-image input register.

Analog inputs: The S7-200 does not update analog inputs from expansion modules as part of the normal scan cycle unless filtering of analog inputs is enabled. An analog filter is provided to allow you to have a more stable signal. You can enable the analog filter for each analog input point.

When analog input filtering is enabled for an analog input; the S7-200 updates that analog input once per scan cycle, performs the filtering function, and stores the filtered value internally. The filtered value is then supplied each time program accesses the analog input.

When analog filtering is not enabled, the S7-200 reads the value of the analog input from expansion modules each time program accesses the analog input.

Analog inputs AIW0 and AIW2 included on the CPU 224XP are updated every scan with the most recent result from the analog-to-digital converter. This converter is an averaging type (sigma-delta) and those values will usually not need software filtering.

iv. Executing the Program

During the execution phase of the scan cycle, the S7-200 executes program, starting with the first instruction and proceeding to the end instruction. The immediate I/O instructions give you immediate access to inputs and outputs during the execution of either the program or an interrupt routine.

If we use subroutines in program, the subroutines are stored as part of the program. The subroutines are executed when they are called by the main program, by another subroutine, or by an interrupt routine. Subroutine nesting depth is 8 from the main and 1 from an interrupt routine.

If we use interrupts in program, the interrupt routines that are associated with the interrupt events are stored as part of the program. The interrupt routines are not executed as part of the normal scan cycle, but are executed when the interrupt event occurs (which could be at any point in the scan cycle).

Local memory is reserved for each of eleven entities: one main, eight subroutine nesting levels when initiated from the main, one interrupt, and one subroutine nesting level when initiated from an interrupt routine. Local memory has a local scope in that it is available only within its associated program entity, and cannot be accessed by the other program entities.

Basic PLC and Elevator Control System

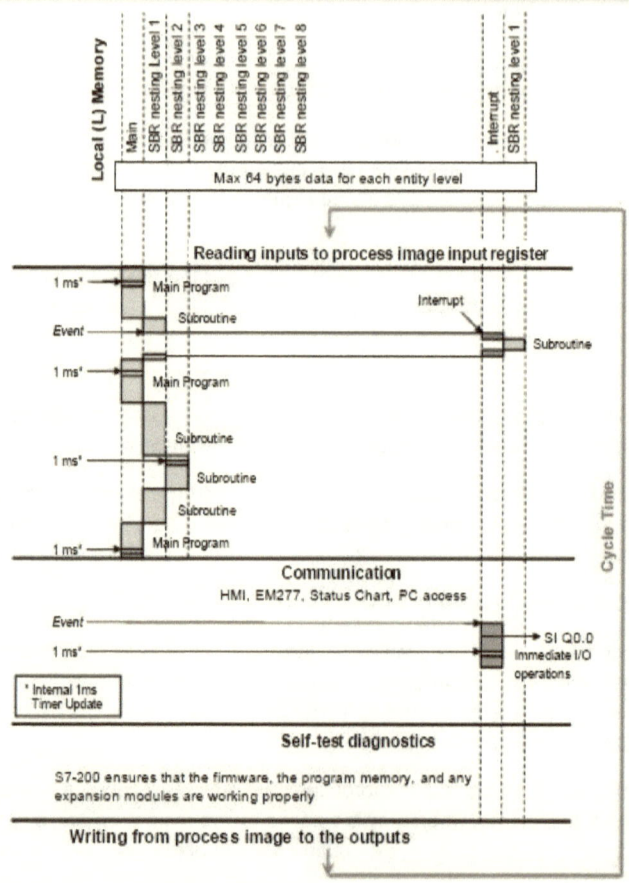

Figure Appendix A.3: Typical Scan Flow

Figure 4-3 depicts the flow of a typical scan including the Local memory usage and two interrupt events, one during the program--execution phase and another during the communications phase of the scan cycle. Subroutines are called by the next higher level, and are executed when called. Interrupt routines are not called; they are a result of an occurrence of the associated interrupt event.

Md. Lutfar Rahman

Basic PLC and Elevator Control System

v. Processing Any Communications Requests

During the message-processing phase of the scan cycle, the S7-200 processes any messages that were received from the communications port or intelligent I/O modules.

vi. Executing the CPU Self-test Diagnostics

During this phase of the scan cycle, the S7-200 checks for proper operation of the CPU and for the status of any expansion modules.

vii. Writing to the Digital Outputs

At the end of every scan cycle, the S7-200 writes the values stored in the process-image output register to the digital outputs. (Analog outputs are updated immediately, independently from the scan cycle.)

Appendix B – About Ladder Logic Diagram

What is Ladder Logic

Ladder logic is a visual programming language used to program PLC's (Programmable Logic Controllers).Ladder logic consists of horizontal Rungs and Instructions embedded between vertical Rails on either side. Rungs house instructions which are referenced by tags or variables. The rails represent the opposing polarity of power rails as shown on an electrical schematic. The rungs and subsequent instructions represent the "load" between the rails.
Rungs and rails are graphical representations of electrical schematics. The real magic behind ladder logic is the instructions, and there are a bunch of them. There are Boolean, Math, Timer, Counter, and Specialty instructions just to name a few. All instructions reference a tag or variable which is an address to a memory location.

What advantage is there to programming in ladder logic?

Let's take a look at the "Figure Appendix B.1" below. On the left is Structured Text and on the right is Ladder Logic. Both programs accomplish the same thing just with different languages. Notice the green vertical rails in the ladder logic example and the green vertical bar in the structured text example. The green let us know that the logic is enabled and running.

Basic PLC and Elevator Control System

```
If Struct_Input THEN;
Struct_Output := 1;
ELSE
Struct_Output := 0;
END_IF;
```
```
Relay Input                Coil Output
Ladder_Input               Ladder_Output
─────] [─────              ─────( )─────
```

Figure Appendix B.1: Structured Text with Ladder Logic

Let's take a look at the rung on the right. Instantly we know the values of the tags are zero. How? That's easy, because if either of them had a value of 1 they would be green just like the rails.

Now we look at the code on the left. It's impossible to determine the input and output tag values without bringing up a watch window and checking the value of the tags. Incidentally all values in both programs are set to 0 in this example.

The" Figure Appendix B.2" below shows the exact same logic, however something is different on the rung example. The instructions are now highlighted in green the same way the rails are in the last example. Both tag values are now set to 1 as indicated by the green highlighting of the instructions.

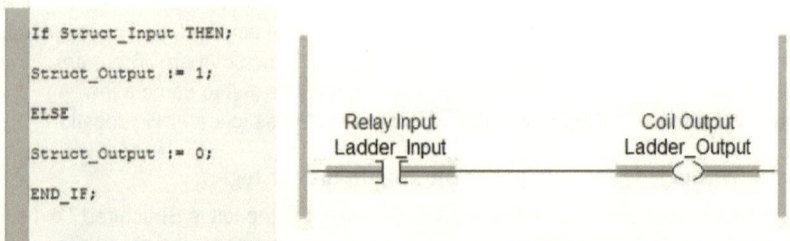

Figure Appendix B.2: Structured Text with Ladder Logic On

In the Structured Text program both values are 1, however the code is just that, code. The code does not provide feedback like the ladder logic example does.

Md. Lutfar Rahman

Basic PLC and Elevator Control System

It's possible to open a watch window to see the value of the two variables, for that matter the same watch window could be opened for the ladder logic program as well. The major difference being is it's unnecessary for the ladder program. Just a quick glance lets us know what is on and what is off.

This kind of feedback is undoubtedly the major advantage of using ladder logic. Ladder logic places all the information us need where us need it. It makes troubleshooting quick and easy, and this functionality isn't limited to just bit instructions. Variables are displayed on instructions as well. Most ladder instructions will show the value of tags in real time. Take a look at the "Equal To" instruction below. If the value of Source A is equal to the value of Source B turn on the output bit. In this case they are not equal and the output bit is off.

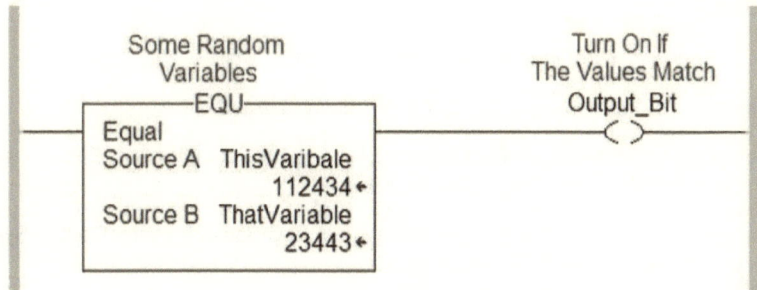

Figure Appendix B.2: the "Equal To" instruction

Now we have been tasked to find out why the variable "Output_Bit" is off. Just a quick glance at the ladder rung and it's easy to determine why the bit is off. The two variables in the comparison instruction are not equal therefore the "Output_Bit" if off. From there we can do a simple cross reference on the variables to discover why the variables are not equal.

Programming ladder logic entails dragging and dropping instructions, rungs and branches. Programming structured text entails knowing the correct syntax. Let's have a look at the structured text examples above. Notice the Semicolons and Colons. This is representative of the Pascal programming language. Instructions are in Blue and tags are in Red. Setting a value is done with the := operator while comparing values is done with just the = operator.

*Ladder logic is scanned the same as a book is read, from top to bottom left to right.

Md. Lutfar Rahman

References

[1] S7-200 Programmable Controller System Manual by Siemens Support Website

[2] David Jin, Sally Lin, Advances In Computer Science, Intelligent System and Environment, Pages 743-748, Chapter The Design of Elevator Control System Based on PLC and Configuration by Shuang Zheng & Fugang Liu

[3] M. Al Mulla, Control of a 4-level Elevator System Using a Programmable Logic Controller, Senior Project, Systems Engineering Department, KFUPM, September 1988

[4] George R. Strakosch, Robert S. Caporale, The Vertical Transportation Handbook, Fourth Edition, Pages 131-259

[5] Roybelt Kuls, Advanced PLC Design Seminars, Shenyang University of Chemical Technology, 2013

[6] PLC Programming of Industrial Automation by Kevin Collins

[7] Beginners of PLC Online Class by Allen-Bradley, Texas

[8] Comprehensive Help from Web Search Engine google.com, Internet Encyclopedia wikipedia.com and Search Engine baidu.com

[9] Sandar Htay, Su Su Yi Mon, Implementation of PLC Based Elevator Control System, Volume-3, Number-2, IJECSE, Pages 91-100

I want morebooks!

Buy your books fast and straightforward online - at one of world's fastest growing online book stores! Environmentally sound due to Print-on-Demand technologies.

Buy your books online at
www.morebooks.shop

Kaufen Sie Ihre Bücher schnell und unkompliziert online – auf einer der am schnellsten wachsenden Buchhandelsplattformen weltweit! Dank Print-On-Demand umwelt- und ressourcenschonend produziert.

Bücher schneller online kaufen
www.morebooks.shop

KS OmniScriptum Publishing
Brivibas gatve 197
LV-1039 Riga, Latvia
Telefax:+371 686 204 55

info@omniscriptum.com
www.omniscriptum.com

www.ingramcontent.com/pod-product-compliance
Lightning Source LLC
Chambersburg PA
CBHW030031250526
45464CB00025B/1239